Jellyfish

by Leighton Taylor
photographs by Norbert Wu

 Lerner Publications Company • Minneapolis, Minnesota

Thanks to our series consultant, Sharyn Fenwick, elementary science/math specialist. Mrs. Fenwick was the winner of the National Science Teachers Association 1991 Distinguished Teaching Award. She also was the recipient of the Presidential Award for Excellence in Math and Science Teaching, representing the state of Minnesota at the elementary level in 1992.

Additional photographs are reproduced through the courtesy of: pp. 4, 12, 16, 20, 21, 23, 28, 30 (both), 41 © 1998 Ben Cropp/Mo Yung Productions; p. 7 © 1998 Lynn Cropp/Mo Yung Productions; pp. 22, 29 © 1998 Peter Parks/Mo Yung Productions; p. 39 © 1998 Bob Cranston/Mo Yung Productions.

Early Bird Nature Books were conceptualized by Ruth Berman and designed by Steve Foley. Series editor is Joelle Riley.

Lerner Publications Company
A division of Lerner Publishing Group
241 First Avenue North
Minneapolis, MN 55401 U.S.A.

Website address: www.lernerbooks.com

Library of Congress Cataloging-in-Publication Data

Taylor, Leighton.
    Jellyfish / by Leighton Taylor ; photographs by Norbert Wu.
      p. cm. — (Early bird nature books)
    Includes index.
    Summary: Describes the life cycle, habitat, behavior, and physical structure of the soft-bodied sea animal that has no brain or bones.
    ISBN 0-8225-3028-7 (lib. bdg. : alk. paper)
    1. Jellyfishes—Juvenile literature. [1. Jellyfishes.] I. Wu, Norbert, ill. II. Title. III. Series.
QL377.S4T39 1998
593.5'3—dc21
                                                   97-22207

Manufactured in the United States of America
3 4 5 6 7 8 – JR – 08 07 06 05 04 03

Contents

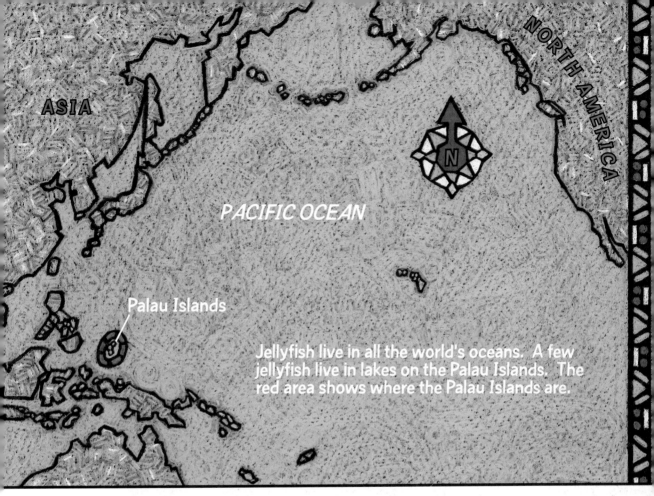

ASIA

NORTH AMERICA

N

PACIFIC OCEAN

Palau Islands

Jellyfish live in all the world's oceans. A few jellyfish live in lakes on the Palau Islands. The red area shows where the Palau Islands are.

Be a Word Detective

Can you find these words as you read about the jellyfish's life? Be a detective and try to figure out what they mean. You can turn to the glossary on page 46 for help.

algae	**planula**	**stinging cells**
bell	**polyp**	**tentacles**
medusa	**predators**	**transparent**
oral lobes	**prey**	

Chapter 1

Sometimes jellyfish look like spaceships. Where do most jellyfish live?

What Is a Jellyfish?

Jellyfish swimming in the sea look like spaceships. Of course, they are not spaceships. But they are not fish, either.

Jellyfish are wonderful water animals. They have bodies that are soft, like jelly. Sometimes jellyfish are called "jellies."

There are hundreds of species, or kinds, of jellies. A few species live in saltwater lakes. But most species of jellyfish live in oceans. Jellyfish can be found in every ocean in the world.

Some jellies live in warm waters.

Near the North Pole, the ocean is very cold. This tiny jellyfish lives in these cold waters.

Many species of jellyfish live in warm parts of the ocean. Some live in cold water. Some jellies swim near the shore. Others swim far away from land.

Jellyfish come in many sizes. One species is as small as a pearl. Another is as long as two school buses.

This jelly is longer than an adult human.

This is a glass jellyfish. It lives in cold, dark waters.

Some jellies are transparent. This means that you can see right through them. Others are mostly white with streaks of color. Some deep-sea jellyfish glow in the dark.

Some jellyfish are mostly white. This jelly has streaks of purple.

Corals are animals who live on strips of rock called reefs. A jelly is swimming over this coral reef.

Jellyfish have thousands of relatives. Corals and sea anemones (uh-NEH-muh-neez) are related to jellyfish.

This jellyfish is swimming over sea anemones. Sea anemones are animals who are related to jellyfish.

A Portuguese man-of-war looks like one animal. But it is really a group of tiny animals who live together.

Another relative has a big name. It is called the Portuguese (POR-choo-geez) man-of-war. It looks like a jellyfish.

Most jellies can sting. Most jellyfish relatives can sting, too. The stings hurt. Never touch an animal who looks like a jellyfish!

Chapter 2

You have something that a jellyfish does not have. What is it?

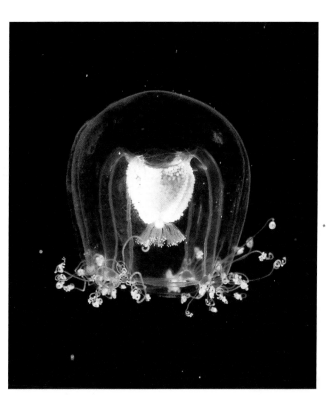

What Jellyfish Can Do

A jellyfish has no brain. But it has nerves. Nerves send messages through the jellyfish's body. A jellyfish can tell many things with its nerves. It can tell when other animals are near. It can tell up from down.

A jellyfish has no bones. But it has a shape. And it has muscles. It uses its muscles for swimming and eating.

The body of a jellyfish is called a bell. Sometimes it is shaped like a bell. Sometimes it is flat. Sometimes it looks like a ball.

The round part of a jelly's body is called a bell.

This jelly's bell is as big as a beach umbrella.

Have you ever filled your mouth with water? Your cheeks puff out. You can spit the water out by squeezing your cheeks.

A jellyfish swims by squeezing its bell. First the jellyfish lets water into its bell. Then the jelly squeezes muscles in its bell. Water rushes out. The jelly shoots ahead.

This jelly is mostly transparent. Its oral lobes look yellow.

A jellyfish has eyes around the bottom of its bell. The eyes can tell the difference between light and dark.

A jellyfish's mouth is beneath its bell. Wavy bands hang down from the mouth. The bands are called oral lobes.

Jellyfish also have tentacles (TEN-tuh-kuhlz). Tentacles stream down from the edge of the bell. Some jellies have short tentacles. Others have long tentacles.

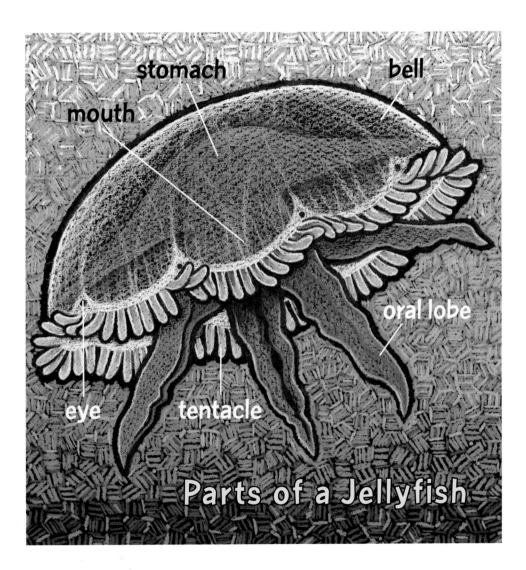

Parts of a Jellyfish

stomach · mouth · bell · oral lobe · eye · tentacle

Most jellyfish can sting. This is a jellyfish called a sea wasp.
It is stinging a fish.

Chapter 3

A fish makes a good meal for a jellyfish. How do jellyfish hunt for food?

Jellyfish Stings

 Most jellyfish are predators (PRED-uh-turz). Predators are animals who hunt and eat other animals. The animals a jellyfish hunts are called its prey. Jellyfish hunt prey such as fish and shrimp.

To hunt, a jellyfish swims near prey. It touches the prey with its tentacles and oral lobes. Then it stings the prey.

Both the tentacles and the oral lobes are covered with stinging cells. The stinging cells pop open when they are touched. A tiny dart comes out of each stinging cell. The darts hook the prey and shoot poison into it.

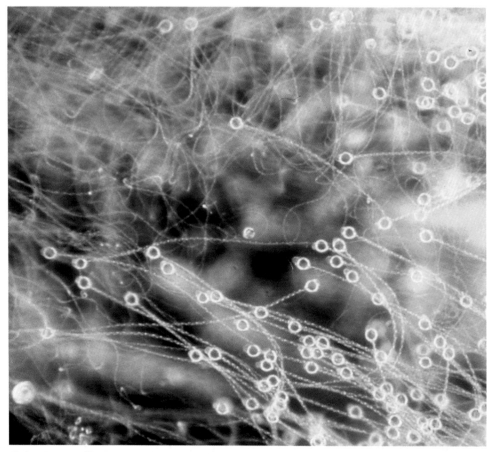

Stinging cells look like circles. They are tiny. This picture makes stinging cells look much bigger than they really are.

A sea wasp is eating a fish.

The poison weakens the prey. The darts hold the prey. It cannot escape. Sometimes the poison kills the prey. Then the jellyfish uses its oral lobes to move the prey to its mouth. It eats the prey. A jellyfish's sting makes it a good predator.

This jelly swam too close to the bottom of the water. A starfish is eating the jelly.

A jellyfish's sting also helps to protect the jelly. A jellyfish stings almost any animal it touches. It stings the animals it eats. But it also stings predators that try to eat it. Most predators do not try to eat jellyfish. They do not want to be stung.

A few predators can eat jellyfish without getting hurt. Some fish can. Some sea turtles, crabs, and snails can, too. These predators have tough mouths and stomachs.

These fish are kelp bass and garibaldi. The bright orange garibaldi has been nibbling on the tentacle of this jellyfish. Ocean sunfish also eat jellyfish.

Chapter 4

The circles in this jelly are the places where the jelly's eggs are made. Where does a jellyfish lay its eggs?

Growing Up

 A jellyfish begins life as an egg. A mother jellyfish makes many eggs in her body. She releases the eggs into the water. The eggs float away from her.

In a few days, each egg becomes a tiny animal. It looks like a worm. This animal is called a planula (PLAN-yoo-luh).

A planula is transparent. It can swim, but mostly it just floats.

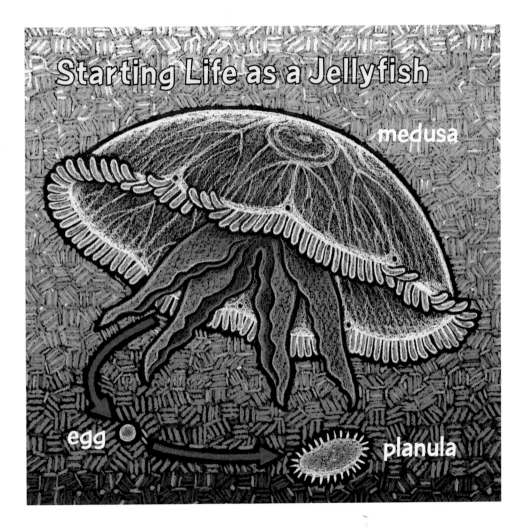

Starting Life as a Jellyfish

medusa

egg

planula

This is a polyp. Polyps are a form of jellyfish. Jellyfish live as polyps for a year or more.

After a few days or weeks, the planula sinks to the bottom of the water. It hooks onto a rock or another part of the bottom. Very soon it grows into a polyp (POL-ip).

A polyp has a mouth and tentacles. But a polyp cannot swim. It is stuck to the bottom. It stings prey animals that swim near. It holds the prey with its tentacles.

Small disks, shaped like saucers, begin to grow from the polyp. By the end of a year, a polyp has many disks. Then the disks break off.

This disk is bumpy. Many disks break away from one polyp.

The disks float for about one week. By the end of a week, each disk has become a medusa (muh-DOO-suh).

Many medusas are made by one polyp. A sea wasp polyp (above) *produces many medusas like this one* (right).

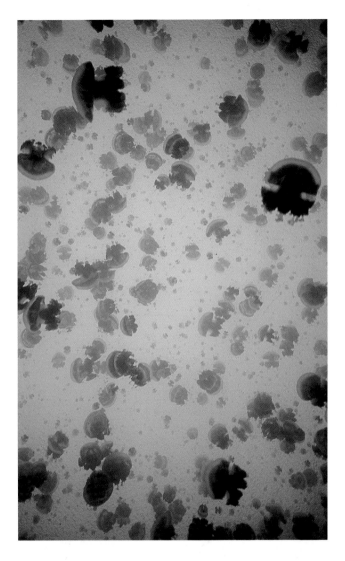

A medusa is the biggest form of jellyfish. These medusas are swimming in the sun.

A medusa is what most people think of as a jellyfish. Some medusas are eaten by predators. Some wash onto beaches and die. Most medusas live for about one year.

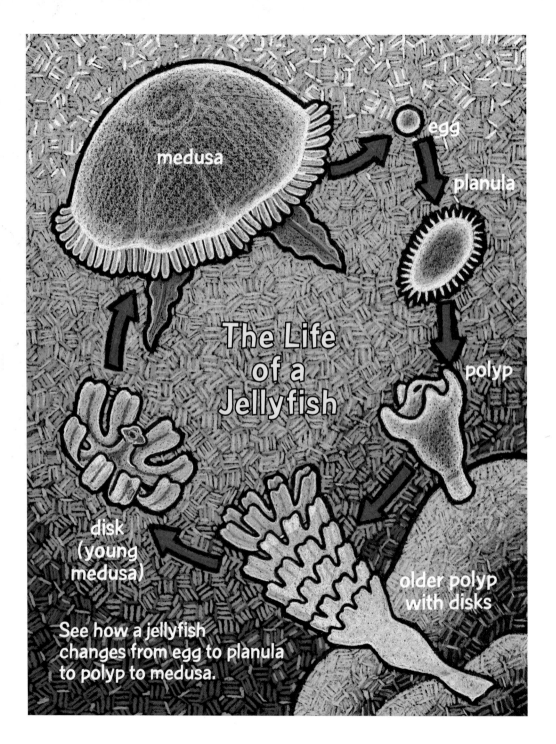

medusa

egg

planula

The Life
of a
Jellyfish

polyp

disk
(young
medusa)

older polyp
with disks

See how a jellyfish
changes from egg to planula
to polyp to medusa.

Chapter 5

The form of jellyfish people see most is the medusa. This is a medusa. Do medusas swim together?

Jellyfish Neighbors

 Most jellyfish swim alone. Sometimes many jellies are found together. They are not traveling together. They just happen to be in the same place.

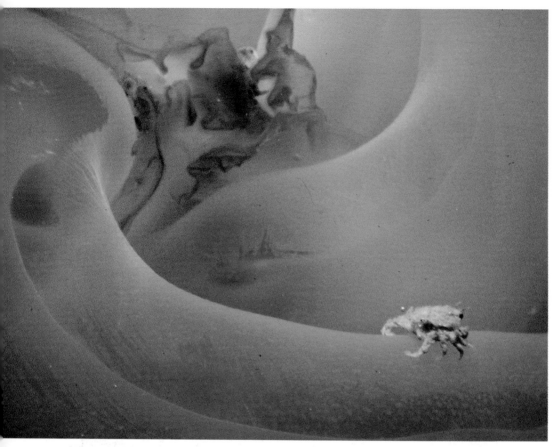

This crab is getting a ride in a jellyfish.

But jellyfish do have neighbors. Some
animals ride inside a jellyfish's bell. Some tiny
crabs do this. Some shrimplike animals do, too.

Animals who ride inside jellyfish share food
the jellyfish catches. Staying inside a jellyfish
also keeps them safe from predators.

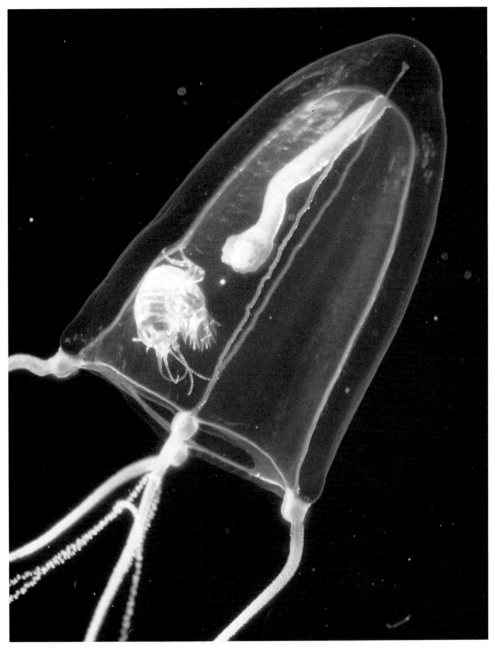

You can see a tiny animal inside the bell of this jelly. The animal is related to shrimps.

Some small fish stay near a jellyfish's tentacles. They swim along with the jellyfish. Predators mostly stay away from them because they are near a jellyfish.

The small fish do not get stung. That is because they have a special coating on their bodies. The coating keeps the jellyfish's stinging cells from opening.

Sometimes small fish swim along with jellies like this one.

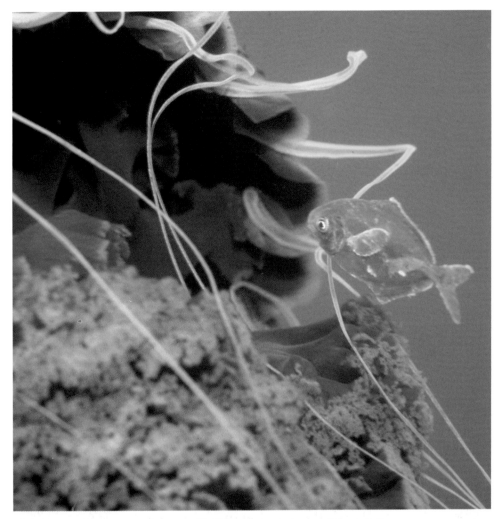

This is a close-up of a small fish swimming with a jelly.

Sometimes a predator does attack these small fish, even though they are close to a jellyfish. The jellyfish may sting the predator. Then the jellyfish and the small fish both eat it.

A jellyfish's bell is also a good place for algae (AL-jee) to live. Algae are tiny plants. They float in water. Sometimes algae live inside the bodies of jellyfish.

Most lakes have freshwater, but some have saltwater. Jellyfish live in this saltwater lake.

These jellies follow the sun as it moves across the sky.

Like all plants, algae need sunlight. Plants use sunlight to make their own food. Some jellyfish in lakes stay in the sunlight. The algae who live with these jellyfish get lots of sun. They can make lots of food. They make so much that the jellyfish gets some, too.

Chapter 6

Sometimes jellyfish and jellyfish relatives wash onto a beach. Are jellyfish on a beach safe to touch?

Watching Jellyfish

 Have you ever seen jellyfish? You can see jellyfish at an aquarium. You can see them in the water at the seashore.

You can also see jellyfish on the beach. Jellyfish on a beach are dead or dying. Look at them, but do not touch them. Even dead jellyfish can sting.

Most jellyfish have a mild sting. A mild sting hurts a person only a little. It gives a person a red rash.

A few jellyfish have a dangerous sting. The sea wasp is a species of jellyfish that swims near the shore of Australia. Its sting can kill people.

This diver is being very careful not to touch a sea wasp.

Sometimes scientists dive underwater to study jellyfish.

People are still learning about jellyfish. One deep-sea jellyfish gave scientists a surprise. A fish attacked the jellyfish. The fish bit into one of the jelly's tentacles. The jellyfish let the tentacle drop off. Then the jellyfish swam away!

No one had ever seen a jellyfish do this. The scientists studied this kind of jellyfish some more. Then they knew they had found a new species of jellyfish.

New species of jellyfish are discovered every year. More new species will be found. Maybe you will help to find them!

Jellyfish are amazing animals. People learn more about jellyfish every year.

On Sharing a Book

As you know, adults greatly influence a child's attitude toward reading. When a child sees you read, or when you share a book with a child, you're sending a message that reading is important. Show the child that reading a book together is important to you. Find a comfortable, quiet place. Turn off the television and limit other distractions, such as telephone calls.

Be prepared to start slowly. Take turns reading parts of this book. Stop and talk about what you're reading. Talk about the photographs. You may find that much of the shared time is spent discussing just a few pages. This discussion time is valuable for both of you, so don't move through the book too quickly. If the child begins to lose interest, stop reading. Continue sharing the book at another time. When you do pick up the book again, be sure to revisit the parts you have already read. Most importantly, enjoy the book!

Be a Vocabulary Detective

You will find a word list on page 5. Words selected for this list are important to the understanding of the topic of this book. Encourage the child to be a word detective and search for the words as you read the book together. Talk about what the words mean and how they are used in the sentence. Do any of these words have more than one meaning? You will find these words defined in a glossary on page 46.

What about Questions?

Use questions to make sure the child understands the information in this book. Here are some suggestions:

> What did this paragraph tell us? What does this picture show? What do you think we'll learn about next? Jellyfish live as planulas, polyps, and medusas. Do you know any other animals who change from one form to another? Is "jellyfish" a good name for a medusa? Why/Why not? How are people different from jellyfish? How are we the same? Could a jellyfish live near you? Why/Why not? How far would you have to travel to see jellyfish? Would you ever touch a jellyfish? Why/Why not? What is your favorite part of the book? Why?

If the child has questions, don't hesitate to respond with questions of your own, such as: What do *you* think? Why? What is it that you don't know? If the child can't remember certain facts, turn to the index.

Introducing the Index

The index is an important learning tool. It helps readers get information quickly without searching throughout the whole book. Turn to the index on page 47. Choose an entry, such as *swimming,* and ask the child to use the index to find out how a jellyfish swims. Repeat this exercise with as many entries as you like. Ask the child to point out the differences between an index and a glossary. (The index helps readers find information quickly, while the glossary tells readers what words mean.)

Where in the World?

Many plants and animals found in the Early Bird Nature Books series live in parts of the world other than the United States. Encourage the child to find the places mentioned in this book on a world map or globe. Take time to talk about climate, terrain, and how you might live in such places.

All the World in Metric!

Although our monetary system is in metric units (based on multiples of 10), the United States is one of the few countries in the world that does not use the metric system of measurement. Here are some conversion activities you and the child can do using a calculator:

WHEN YOU KNOW:	MULTIPLY BY:	TO FIND:
miles	1.609	kilometers
feet	0.3048	meters
inches	2.54	centimeters
gallons	3.787	liters
pounds	0.454	kilograms

Activities

Pretend you are a scientist, and you are planning a trip to a place where you can study jellyfish. Find the place on a map or globe. How will you get there? What will you take? Who will go with you? What will you see? Write a story about your trip.

A medusa is a form of a jellyfish. But Medusa was also a person in Greek legend. Go to the library and find out about Medusa of ancient Greece.

Color or paint a picture of a medusa. Some medusas are white with streaks of color. The streaks can be purple, silver, red, or blue. Make your medusa your favorite color.

Glossary

algae (AL-jee)—small, rootless plants that grow in water

bell—the main part of a jellyfish's body

medusa (muh-DOO-suh)—the biggest form of a jellyfish

oral lobes—wavy bands that hang down from a jellyfish's mouth

planula (PLAN-yoo-luh)—the form of a jellyfish that looks like a tiny worm

polyp (POL-ip)—the form of a jellyfish that is stuck to the ground

predators (PRED-uh-turz)—animals who hunt and eat other animals

prey—animals who are hunted and eaten by other animals

stinging cells—tiny parts of a jellyfish that fire poisonous darts

tentacles (TEN-tuh-kuhlz)—stringy parts that hang down from the edges of a jellyfish

transparent—clear

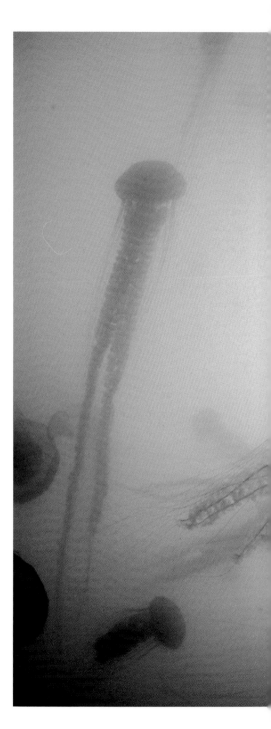

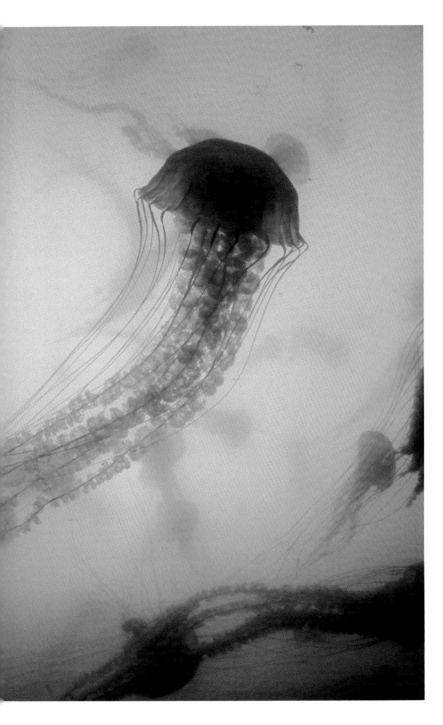

Index

Pages listed in **bold** type refer to photographs.

About the Author

Leighton Taylor is a marine biologist who began studying the sea while fishing as a small boy in California. He went to graduate school in Hawaii. Hawaii's warm water, bright fish, and coral reefs convinced him to spend his life studying and writing about the animals who live in the sea. He earned a Ph.D. degree at Scripps Institution of Oceanography. He loves to dive and has made many expeditions in the Pacific Ocean, the Indian Ocean, and the Caribbean Sea. He has discovered and named several new species of sharks, including the deep-sea Megamouth shark.

About the Photographer

Norbert Wu's photography has appeared in numerous books, films, and magazines, including *Audubon, Harper's, International Wildlife, Le Figaro, National Geographic, Omni, Outside, Smithsonian,* and the covers of *GEO, Natural History, Time,* and *Terre Sauvage.* The author and photographer of several books on wildlife and photography, his photographic library of marine and topside wildlife is one of the most comprehensive in the world. His recent projects include television filming for National Geographic Television, Survival Anglia, and PBS.